The Bee Spotter's Guide

The Bee Spotter's Guide

DAVE GOULSON

Illustrated by
Ella Sienna

Published by National Trust Books
An imprint of HarperCollins Publishers,
1 London Bridge Street London SE1 9GF
www.harpercollins.co.uk

HarperCollins Publishers,
Macken House,
39/40 Mayor Street Upper,
Dublin 1, D01 C9W8, Ireland

First published 2026

© National Trust Books 2026
Text © Dave Goulson
Illustrations © Ella Sienna

ISBN 978-0-00-878131-6
10 9 8 7 6 5 4 3 2 1

All rights reserved. No part of this publication may be reproduced, stored in a retrieval system, or transmitted, in any form or by any means, electronic, mechanical, photocopying, recording or otherwise without the prior permission in writing of the publisher and copyright owners.

Without limiting the exclusive rights of any author, contributor or the publisher of this publication, any unauthorised use of this publication to train generative artificial intelligence (AI) technologies is expressly prohibited. HarperCollins also exercise their rights under Article 4(3) of the Digital Single Market Directive 2019/790 and expressly reserve this publication from the text and data mining exception.

The contents of this publication are believed correct at the time of printing. Nevertheless, the publisher can accept no responsibility for errors or omissions, changes in the detail given or for any expense or loss thereby caused.
A catalogue record for this book is available from the British Library.
Printed by Multivista Global Pvt. Ltd, India

If you would like to comment on any aspect of this book, please contact us at the above address or national.trust@harpercollins.co.uk

National Trust publications are available at National Trust shops or online at nationaltrustbooks.co.uk

Contents

Introduction   6

A–Z of Bees   10–101

   Features

   Bee Life Cycles 26

   Bee Enemies 46

   Best Places to See Bees 66

   Gardening for Bees 86

Bee mimics   102–105

Recommended guides   106

Index   107

Introduction

Bees evolved about 120 million years ago, at a time when dinosaurs roamed the Earth. Their ancestor was a wasp that turned vegan, switching from preying on insects to feeding its offspring with pollen. It proved to be a successful strategy because, in spite of the meteor collision 65 million years ago that exterminated the dinosaurs and many other creatures, the bees not only survived but proliferated. Today, there are over 20,000 known species, about 270 of which can be found in Britain and Ireland.

Few people are aware of the diversity of bees. Everybody has heard of the honey bee, but when asked to draw one, many people will come up with a rounded, furry creature with yellow and black stripes – a bumblebee. In these pages you will discover that there are many types of bee: mason bees, mining bees, cuckoo bees, carder bees, blood bees, leafcutter bees and more. Some, such as the scissor bee, are just a few millimetres in length and can be easily overlooked. Others, such as the huge, droning queens of buff-tailed bumblebees, are impossible to miss.

Bees can be found almost everywhere as long as there are flowers. Many species can be found in gardens and urban parks, so everyone has the opportunity to encounter them. When bees are preoccupied with visiting flowers, you will find that you can easily get close to them. With practice, you will notice that different bees have different flower preferences. In part this relates to their size, shape and the length of their tongues. Small bees can fit into tiny flowers, while those with long tongues, such as the garden bumblebee, can reach the nectar in deep, tubular flowers such as foxgloves.

When out spotting bees, keep an eye out for insects that pretend to be bees. There are many, including beetles, moths and particularly flies, all of which try to avoid being eaten by pretending to be a creature

Buff-tailed bumblebee (see page 20).

Dark-edged bee-fly (see page 102).

with a sting. These mimics include some of our most splendid insects, such as bee hawk-moths, bee flies (see page 102) and bumblebee hoverflies (see page 104).

Take the time to sit and watch bees for a little while and you will soon observe interesting behaviours. Find a patch of comfrey or monkshood and you will begin to see bumblebees nectar-robbing by biting holes in the back of the flowers. Look out for nesting behaviour; bumblebee queens fly low to the ground in early spring looking for pre-existing holes, while mining bees dig their own. Mason and leafcutter bees prefer to nest above ground in hollow stems or in beetle holes in dead wood, and will readily adopt commercial or homemade 'bee hotels'. Some even nest in empty snail shells. Wherever they make their home, you may see cuckoo bees skulking about, waiting for the chance to sneak into the nest and lay their own eggs. All of this and much more can be seen in an urban garden, though you may want to hunt further afield in flower-rich meadows, heaths and woodland glades if you wish to see more unusual species.

Red-tailed mason bee (see page 78).

Ashy mining bee
Andrena cineraria

There are 67 species of mining bee (*Andrena*) in the UK. All are solitary bees – bees that don't live in colonies – with females burrowing into the ground to nest. They often close their nest entrances at night or during rain to protect it. The nests are usually less than 50 centimetres, but ones as deep as 5 metres have been reported. Some mining bees can be tricky to identify, but the ashy mining bee is one of the most distinctive. It is relatively large and lives up to its name, as it is ash-coloured with the female sporting a shiny black abdomen. They produce just one generation per year, with adults on the wing from March to June.

Ashy mining bees can be found in all sorts of habitats, including arable farmland, abandoned brownfield sites and gardens. Although they are solitary bees, their nest burrows are often found in large clusters in bare soil, farm track edges, short rabbit-grazed turf and garden lawns. Ashy mining bees are widespread in England, Wales and Ireland, being most abundant in central and western England, but are rare in Scotland. This species visits a broad range of flowers, including willow catkins, hawthorn, blackthorn, hogweed and charlock.

I SPOTTED THIS BEE

AT

ON

Bare-saddled blood bee
Sphecodes ephippius

These small but pretty bees are sparsely haired with a bright red abdomen. Blood bees are cuckoos, with the bare-saddled blood bee targeting several species of furrow bee. The female blood bee may be diminutive, but she is brutal. She will enter the nest of a host, forcing her way past the owner if necessary. She then opens a host cell, kills the offspring with her sickle-shaped jaws, and replaces it with her own egg.

Female bare-saddled blood bees are on the wing from April to September, with the males flying later, from July to October. The females are most often seen near host nests, while males frequently graze on flowers, favouring dandelions and umbellifers such as hogweed.

This is a species of open, sunny habitats, often found on brownfield sites. It is widespread and sometimes common in much of England and Wales, but very scarce in Ireland and Scotland.

I SPOTTED THIS BEE

AT

ON

Bilberry bumblebee
Bombus monticola

With its huge red bottom and black and yellow stripes, the bilberry bumblebee is a strong contender for the title of the UK's most colourful and beautiful bee, and is one that is very easy to identify. It is also unusual in that it is strongly associated with upland heaths and moors where bilberries grow, with strongholds in Scotland, the Pennines, Wales and Dartmoor; a bilberry bumblebee hunt is the perfect excuse for a hike in the hills. Queen bilberry bumblebees feed mainly on mountain willows and sallows, then move on to bilberry flowers in mid-spring.

Once bilberry finishes flowering, the worker bees seek out flowery grasslands with clovers and bird's-foot trefoil, and are often forced to descend to lower altitudes to seek them out. Later in the summer they switch back to feeding on heathers in the mountains.

Although sadly declining in the UK, bilberry bumblebees were first recorded in Ireland in 1974, possibly having been blown over from Wales, and seem to be faring well there.

I SPOTTED THIS BEE

AT

ON

Blue mason bee
Osmia caerulescens

These stocky, dark bees barely earn their name, with just the slightest hint of metallic blue visible in sunshine. They commonly nest in bee hotels, or otherwise will exploit holes in dead trees or crevices in walls. This preference presumably explains why this species is most common in gardens and parks near artificial structures. Cells for offspring are constructed from chewed-up leaves, the final plug being neat, smooth, greenish and concave, in contrast to the messy mud plugs of the more common red mason bees. Blue mason bees are one of very few regular residents of bee hotels that produce two generations per year, with adults on the wing from late April to August.

The flowers it favours include bird's-foot trefoil, knapweed, white clover, ground ivy, catmint and brambles. Blue mason bees are widespread in England and Wales but very scarce in Scotland and absent from Ireland.

I SPOTTED THIS BEE

AT ....

ON ....

Bryony mining bee
Andrena florea

The best way to see this bee is to simply watch a patch of flowering white bryony (*Bryonia dioica*); look for an attractive little bee with two red stripes near the front of its abdomen. The female bryony mining bee is very fussy about the flowers she will visit, and she only collects pollen from white bryony, gathering the bright yellow pollen on her hairy hind legs. Males also favour white bryony, but both males and females will drink nectar from other flowers, particularly brambles, raspberry and dropwort.

Bryony mining bees are only found in the south-east of England and East Anglia, and then mainly on sandy soils where they dig their nests. They can be quite common in heathland and along woodland rides, where large groups of nests can sometimes be found.

Bryony mining bees produce one generation a year, timing the adult flight period to coincide with the flowering of their food plant from late June to mid-July.

I SPOTTED THIS BEE

AT

ON

Buff-tailed bumblebee
Bombus terrestris

Our most common bumblebee, found throughout the UK except for some Scottish Isles, the buff-tailed bumblebee is the archetypal bee: big and furry with yellow and black stripes. The queens are huge and are easy to spot in early spring when they emerge from hibernation. They fly close to the ground, with a characteristic zig-zag flight, looking for a hole in which to nest. While the queens have a distinctive buff-coloured tail, the smaller workers have a white tail, usually with a slight buff tinge. They are easily confused with the white-tailed bumblebee, which has a pure white tail and slightly paler yellow stripes.

The buff-tailed bumblebee (show here on pussy willow catkins) is particularly abundant in parks and gardens, and in summer can often be seen in considerable numbers on garden plants, such as lavender, catmint and borage. In cities, some buff-tailed bumblebees remain active all year round, taking advantage of the heat island effect and exotic, winter-flowering shrubs.

The buff-tailed bumblebee was voted the UK's favourite insect in 2015.

I SPOTTED THIS BEE

AT ....

ON ....

Chocolate mining bee
Andrena scotica

Neither its common nor Latin name are particularly helpful in describing this bee; disappointingly, it does not mine chocolate, and it is far more common in England and Wales than in Scotland. The common name presumably refers to its colour, although I would describe it as reddish-brown rather than chocolate-coloured.

About the size of a honey bee, the chocolate mining bee is on the wing from mid-March to July, and can be found in diverse habitats including gardens and farmland. It commonly feeds on willows, blackthorn (pictured) and hawthorn, and assists in oilseed rape pollination in farmland.

Chocolate mining bees appear to have a partial second brood, with a few adults turning up in late summer and autumn, often seen on ivy flowers. Nest burrows can be hard to spot as they are often hidden among vegetation, usually on sunny banks and slopes. Unusually, several females will often share a common nest entrance, each having their own side-burrow inside.

I SPOTTED THIS BEE

AT

ON

Common carder bumblebee
Bombus pascuorum

If you see a bumblebee that is brown all over, the chances are very high that it is a common carder. This scruffily nondescript but charming bee is one of the most common, found throughout the UK. It will turn up in almost any habitat that has flowers, and is particularly common in gardens, parks, woodland edges and farmland hedgerows. Common carders have a longer tongue than most of our native bumblebees and use it to drink nectar from both shallow and deep flowers. They have very varied tastes, feeding on a huge range of native and garden plants, with favourites that include lavender, catmint, dead-nettle, teasel, verbena (pictured), comfrey, brambles, thistles and clovers.

Nests are usually on the soil surface, under leaf-litter or in tussocky grass. The nesting cycle of this species is longer than any other of our bumblebees, with queens founding nests in March, and the nests producing new queens and males as late as September or October. A few workers can be seen on the wing until the first harsh frosts of winter finish them off.

I SPOTTED THIS BEE

AT

ON

Honey bee (clockwise from top right): queen, drone and worker.

Bee Life Cycles

The earliest bees were solitary creatures, as are most insects. The female bee would build her own small nest, provisioning it with pollen, perhaps stuck together with sticky nectar. One or more eggs would be laid, and then she would seal up the nest and leave her offspring to look after themselves. The large majority of the world's 20,000 bee species still do this, including most British species, but a few have evolved more social behaviour.

Some bees, such as the common furrow bee (see page 30), display behaviour that is known as eusocial: the females that emerge in spring first rear a batch of half a dozen daughters, who stay with their mother and help her rear a second brood rather than trying to rear their own offspring. This second batch contains both males and females, the latter going on to hibernate and build their own nests the following spring. The first brood of females that do not themselves breed are known as workers, and the nest-founding females are queens.

Bumblebees take this a step further, rearing multiple batches of workers through the spring, until there may be several hundred of them in the nest. In late spring or summer, the nest switches to producing new queens and males, which then fly off, mate and then the newly mated young queens burrow into the soil to sleep until next spring. The workers and the old queen, as well as the males, all die off as the summer ends.

Honey bees (see page 54) take social living to the next level. Their colonies contain many tens of thousands of workers, with a single queen laying up to 2,000 eggs per day in spring. They make honey in the warm days of summer as a food supply for the winter, when there will be few flowers, and it is too cold to fly. In this way the colony does not need to hibernate; instead the bees huddle together around the queen, shivering to generate heat, and feeding on honey as they need it. All being well, a queen can live for several years. If the queen dies or becomes weak, or the colony becomes overcrowded, new queens can be reared by feeding bee grubs on royal jelly. Colonies can split, with a queen and several thousand workers departing to find new accommodation, a process known as 'swarming'. As a result, queen honey bees never have to singlehandedly build a new nest from scratch in the way other bees do.

An interesting aspect of social behaviour in bees is that there is a division of labour. In bumblebees, small workers stay in the nest and look after the brood, while larger workers go out in search of food. In honey bees, it is the young workers that look after the offspring, and older bees that go foraging for food. It is thought that bees get better at particular jobs with practice, so it makes sense for each to specialise in doing one thing well.

Honey bee (see page 54).

Common furrow bee

Lasioglossum calceatum

This bee has a very long flight season, from March to October, vying with the ivy bee (see page 56) to be one of the last bees on the wing in autumn.

Common furrow bees are eusocial in southern UK, meaning that there is a queen and worker bees, similar to the arrangement used by bumblebees (see page 27). The queen digs a nest – a burrow in the ground – singlehandedly in spring, rearing a small group of four or five daughter workers who then help her rear a batch of new queens and males in summer. Nests are usually in bare or sparsely vegetated ground in a sunny situation.

The common furrow bee's favourite flowers include dandelion, bramble, ragwort and buttercup (pictured), though they are not fussy. Males will sleep together in clusters on flower heads at night. This species is widespread and often common throughout the British Isles but is rare in Ireland. It is found in almost every flowery habitat, including urban areas.

I SPOTTED THIS BEE

AT

ON

Common mourning bee

Melecta albifrons

A large, drab, dark-coloured bee with distinctive white spots of fur along its sides, the common mourning bee is a cuckoo of the hairy-footed flower bee (see page 52). Females are most likely to be seen lurking suspiciously near nests of their host species, awaiting their chance to sneak in unnoticed and lay an egg. As is typical of most cuckoo bees, the mourning bee grub first murders the host larva using specially elongated jaws, and then consumes its food supplies.

Common mourning bees are widespread in England, but rare in Wales and absent from Ireland and Scotland. Their flight season matches that of their host, from March to early June.

This species is often found in gardens, and you can increase the chances of spotting one by making a bee hotel for their hosts from clay, and by growing the host's favourite flowers; particularly lungwort. Mourning bees rarely visit lungwort themselves, but seem to prefer blossoming fruit trees.

I SPOTTED THIS BEE

AT

ON

Common yellow-face bee

Hylaeus communis

These handsome, little, wasp-like bees are easily recognised by their distinctive bright yellow facial markings. This species is common in most of England and Wales but very scarce in Scotland and Ireland.

The common yellow-face bee naturally nests in hollow plant stems, such as dead brambles, but will happily adopt bee hotels as long as they have some small holes (less than 6 millimetres in diameter). Nest cells are separated by a cellophane-like substance secreted by the female bee. Yellow-face bees are unusual in that they don't carry pollen on their legs or in a pollen brush, but instead swallow it and then regurgitate it once in the nest.

This species is frequently seen in gardens. It may also be found in scrub, open woodland and flower-rich grasslands. The favourite flowers of the common yellow-face bee include hogweed, wild carrot and other umbellifers, brambles, as well as plants from the daisy family, including thistles and ragwort. They fly from late May to mid-September.

I SPOTTED THIS BEE

AT

ON

Dull-vented sharp-tail bee

Coelioxys elongata

The sharp-tailed bee is a prime example of a cleptoparasite – insects that steal food or resources from other species. The female will target the nests of leafcutter bees and use her pointed abdomen to stab through the nest's leafy plug to inject an egg. The young sharp-tailed bee grub has vicious-looking jaws which it uses to kill the host's offspring before consuming the nest's food store at its leisure.

The dull-vented sharp-tailed bee is widespread in southern and eastern England, but is never common. It is extremely scarce in Wales, Scotland and Ireland. It times its adult flight season to coincide with its main host, Willughby's leafcutter bee (see page 96), and is on the wing from June to August.

The most likely place to spot these bees is hanging around bee hotels where their host species is nesting. Aside from gardens, you can also find this species on heaths, dunes and brownfield sites.

I SPOTTED THIS BEE

AT

ON

Early bumblebee
Bombus pratorum

The development of a bee's colony consists of a series of stages, from its establishment, growth and reproduction, to its ultimate decline. The common name of this bumblebee derives from its very short colony cycle, with queens emerging from hibernation in March, and founding nests that produce new queens and males in late May. The males are particularly adorable, being fluffy and brightly coloured with lemon-yellow bands and a rusty red bottom.

This species is widespread in the UK and is common in gardens. The early bumblebee is one of our smaller bumblebees, and very timid. Even if you accidentally disturb their nests – which are often in compost heaps or under stones in garden rockeries – they are not inclined to sting.

Queen early bumblebees love sallow catkins and flowering currant, while the workers feed on many flowers, including hardy geranium (pictured), cotoneaster, raspberry and brambles. There is a partial second generation with some new queens immediately nesting in June, so small numbers of early bumblebees can be seen through July and August.

I SPOTTED THIS BEE

AT

ON

Garden bumblebee
Bombus hortorum

The garden bumblebee is notable for having one of the longest tongues of all British bee species – up to 15 millimetres in length, which along with its elongated, 'horsey' face enables it to feed on nectar that is beyond the reach of other pollinators. Therefore, if you spot a yellow and black striped bumblebee visiting a foxglove (pictured), you can be fairly sure it is a garden bumblebee. Other flowers they favour include bluebells, comfrey, dead-nettles and red clover.

When collecting pollen – which like other bumblebees and honey bees they carry in baskets on their hind legs – garden bumblebees have a strong preference for red clover. Garden bumblebees are important in the veg patch too, as one of the main pollinators of broad and runner beans.

Although a widespread species found throughout the UK, garden bumblebees are rarely abundant. As the name suggests, they are commonly found in gardens, but also in woodland edges, hedgerows, brownfield sites and flower-rich meadows.

I SPOTTED THIS BEE

AT ....

ON ....

Gold-tailed melitta
Melitta haemorrhoidalis

Clothed in fluffy yellowish hairs, with orange hind legs and tail, the gold-tailed melitta is a pretty bee. Like the bryony mining bee (see page 18) the gold-tailed melitta is very particular in its flower choices, collecting pollen only from harebells (pictured) and other bellflowers (*Campanula* species), including garden varieties. Therefore, it is only found in places with an abundance of these plants, notably unimproved grasslands such as chalk downland, heaths and – rarely – woodland edges and rides on chalky soils.

This lovely bee may also appear in gardens and parks as long as there are plenty of bellflowers. It will often rest overnight in cute bundles of several individuals huddled inside the flower heads.

On the wing from July to September, the gold-tailed melitta is mainly found in the south and east of England, although there are a few records from south-east Scotland and central Wales. It is absent from Ireland.

I SPOTTED THIS BEE

AT

ON

Gooden's nomad bee
Nomada goodeniana

Nomad bees are a group of slender, wasp-like bees, which are largely hairless and often feature prominent yellow stripes. All of them are cuckoos of ground-nesting mining bees. There are 34 types of nomad bee in the UK, and each tends to be specific to one or two host species. Gooden's nomad mainly attacks nests of the buffish mining bee. The presence of hovering female nomad bees is often the easiest way to find nests of the host species. They wait for the nest owner to leave in search of food and then sneak in to lay an egg. Like the cuckoo bird, the nomad bee grub kills the mining bee grub and then consumes its food.

Gooden's nomad is found throughout the UK, being common in the south. It flies from April to June, and can be found in a broad range of habitats, including gardens.

I SPOTTED THIS BEE

AT

ON

Bee Enemies

Despite their defensive sting, bees face threats from many predators that aspire to eat them, their brood or their honey. Bee-eaters – a group of birds that specialises in catching bees in mid-air – may be rare in Britain, but great tits commonly eat bumblebees, catching them on flowers and pecking out the sting before scooping out the tasty flight muscles. Foraging for nectar and pollen is a dangerous business for bees, as they may become ensnared in spiders' webs or be snatched by crab spiders – camouflaged sit-and-wait predators that lurk on flowers. Conopid flies are specialist parasitoids of bumblebees. They will stab an egg into an unsuspecting bee while it visits a flower, and thereby condemn the poor bee to death while the fly's larva consumes it from the inside.

Bee nests also face many threats. Cuckoo bees may sneak into nests of solitary bees and lay their own eggs, which hatch into murderous grubs that kill the young of the host bee. Bee flies and ruby-tailed wasps have a similar strategy. Some parasitoid wasps have enormously long egg-laying tubes, which they use to inject their eggs into solitary bee grubs when deep within their nest. Cuckoos and parasites are sometimes so numerous that it is a wonder that any host bees survive.

Bumblebee nests are also at risk from cuckoo bumblebees; there are six species in the UK that specialise in invading established nests, killing the queen and enslaving the workforce. Wax moths may also sneak in and lay batches of eggs, the resulting larvae tunnelling through the bumblebee brood

Badgers will excavate nests and consume everything.

Great tits peck out a bumblebee's sting before devouring its flight muscles.

in protective silken tubes, eventually reducing the nest to a crumbling ruin. Most catastrophically, badgers will excavate bumblebee nests and consume everything; the brood, the small honey stores they make, and even the adult bees, leaving nothing but a crater in the ground.

Honey bee nests may be savaged by honey buzzards, which eat both grubs and honey, and seem immune to the sting of the adults, but parasitic mites are probably a greater threat. In particular, the *Varroa* mite, originally from Asia but carelessly spread around the world by humans, is a serious problem, as the European honey bee has little resistance to this parasite or to the viral diseases that it transmits when sucking the bees' blood.

Other novel threats to bees are on the horizon; in particular, the Asian hornet, accidentally shipped to France from China, has spread across much of Europe and has started to turn up in southern England in recent years. It feeds on flying insects but has a particular taste for bees and can wipe out honey bee populations by returning repeatedly to the same colony, slowly depleting the workforce.

The conopid fly is a specialist parasitoid of bumblebees.

Great yellow bumblebee

Bombus distinguendus

One of the UK's most splendid bees, the great yellow bumblebee is also one of the rarest. It was once widespread in the UK, though always more common in the north, but today you will have to travel to the far north coast of mainland Scotland, Orkney or the Outer Hebrides to see one of these magnificent insects; rest assured it is well worth the journey. Their last strongholds are windswept coastal areas, particularly the machair grasslands on the Uists, a beautiful, flower-rich habitat growing on wind-blown shell sand that supports similar flowers to those found on chalk downland further south.

Great yellow bumblebees are easy to identify; their name is apt, for they are large and entirely yellow aside from a black stripe across the thorax. They have long tongues and favour deep, tubular flowers, such as red clover, tufted vetch, kidney vetch, knapweed (pictured) and yellow rattle. They often make their nests in rabbit and rodent burrows in sandy soils.

I SPOTTED THIS BEE

AT

ON

Hairy-footed flower bee

Anthophora plumipes

One of my favourite bees, their emergence in early March is a welcome harbinger of spring. The fluffy brown males emerge first, followed a week or two later by the females, jet black with a bright yellow pollen brush on their hind legs. Their common name derives from the mid-legs of the male, which sport a tuft of long hairs used to stroke the face of the female during courtship. They have a distinctive, hovering and darting flight quite different to most other bees.

If you live in central or southern England it is worth planting lungwort, one of their favourites, just to attract them. They also like to visit comfrey, ground ivy, bugle, green alkanet and primrose (pictured).

Aside from gardens and parks, this bee can be found in scrub and woodland edges, with flight continuing until June. They dig nest burrows into the soft mortar of old walls and cliff faces, and they can be tempted into bee hotels made from clay.

I SPOTTED THIS BEE

AT

ON

Honey bee
Apis mellifera

This is the domestic bee, cared for in hives throughout the world; it provides us with delicious honey and plays an important role in pollinating many crops.

Thanks to domestication, honey bees have become the most globally widespread bee species. Their large colonies are unusual; they contain as many as 80,000 worker bees, along with the queen, who lays several thousand eggs per day in spring and summer. To support these colonies, honey bees have evolved a 'waggle dance', whereby scout bees can communicate the direction and distance of good patches of flowers – which can be several kilometres from the hive – to their nestmates. They build up stores of honey so that colonies can survive the winter when it is too cold to feed.

Honey bees have very varied tastes, visiting a huge range of wild and cultivated plants. It is shown here on white clover. Unfortunately, the abundance of domestic honey bees often leads to competition with wild pollinators for nectar and pollen.

I SPOTTED THIS BEE

AT ....

ON ....

Ivy bee
Colletes hederae

The ivy bee is the last bee to appear on the wing each year. The adults emerge in September to catch the autumn flowering of ivy, which they feed on almost exclusively. The nests become a frenzy of activity as males scramble to win a mate, with newly emerged females being surrounded by a ball of frantic suitors.

Peculiarly, this conspicuous, stripy bee was not described by scientists until 1993, although it was common across Europe. It was first recorded in the UK in 2001, when it was discovered in Dorset. The ivy bee has since spread through England and Wales, though it has not yet reached Scotland or Ireland. It is now very common in the south, forming large groupings of thousands of nests. In nature these nests are found in steep, sunny banks and cliffs, but ivy bees will also nest in garden lawns and flower beds.

I SPOTTED THIS BEE

AT ....

ON ....

Large-headed resin bee
Heriades truncorum

Once regarded as a very rare bee, this species has become quite common in south-east England, especially in Surrey and Sussex, though absent from the rest of the UK. It may be benefitting from the popularity of bee hotels – this species will happily occupy hotels, as long as they include small holes (4–5 millimetres in diameter). In nature it nests in hollow plant stems and holes in dead wood. Inside the nests, the cells are separated by pine resin collected by the bees, and the final plug is strengthened with tiny pebbles.

This species flies from June to September, timed to coincide with the flowering of its favourite plant, ragwort. It will also visit other yellow flowers in the daisy family (Asteraceae), such as hawkweed and fleabane.

Large-headed resin bees can be found in a broad range of habitats wherever ragwort is abundant, including gardens, brownfield sites and overgrazed pastures.

I SPOTTED THIS BEE

AT

ON

Long-horned bee
Eucera longicornis

One of our most charming bees, the male long-horned bee is unmistakable due to its ridiculously long, curved antennae. Sadly this species has declined greatly in recent years and is now found at only a handful of sites, mostly coastal, in southern England and Wales. It has never been recorded in Scotland or Ireland.

Long-horned bees are mainly found on clay soils, particularly on unimproved grasslands rich in meadow vetchling and other plants from the pea family (Fabaceae), such as bird's-foot trefoil (pictured), kidney vetch and clovers. They also occur in woodland rides and clearings.

The females nest by burrowing into clay banks, and there is a thriving population inside the perimeter of Gatwick Airport, nesting in an old clay spoil heap. As with most ground-nesting bees, sizeable groupings of nests can form.

In Europe, male long-horned bees pollinate bee orchids, the flowers of which mimic the appearance and smell of the female bee, but this has not been recorded in the UK. They fly between mid-May and July.

I SPOTTED THIS BEE

AT

ON

Orange-legged furrow bee

Halictus rubicundus

A small, dark bee with conspicuous thin yellow bands on the abdomen, and distinctive orange hind legs in the female. This species has a very long flight period, from March to October.

The orange-legged furrow bee is eusocial in southern England (see page 27), with queens emerging in early spring to excavate a nest and rear workers – female bees that are identical to the queen in appearance but do not usually lay eggs. In late summer, nests produce new queens and males. In contrast, further north in the UK, this species is solitary.

The orange-legged furrow bee is widespread and fairly common throughout the UK, found in a great range of flower-rich habitats, including gardens and upland meadows. It is shown here on devil's-bit scabious. Nests are in burrows in the ground, usually on south-facing slopes, and often found in groups.

I SPOTTED THIS BEE

AT

ON

Pantaloon bee
Dasypoda hirtipes

The name of this beautiful bee derives from the very long orange hairs on the hind legs of the females, which give them the appearance of wearing baggy orange trousers. They are even more striking when the hairs are packed with a full load of bright yellow pollen.

This bee is most likely to be seen on sandy coastal sites, particularly the south and east coasts of England from Dorset to Norfolk, as well as dunes, heaths, salt marshes and weedy brownfield sites. They can also be found close to central London along the Thames estuary.

Nest burrows can be exceptionally deep, sometimes more than 60 centimetres. Pollen is gathered only from members of the daisy family (Asteraceae), particularly those with yellow flowers, including cat's ear, hawkbits, ragwort and fleabane.

Although localised, pantaloon bees can be common where they occur, often nesting in large groups along track edges. They are on the wing from late June to August.

I SPOTTED THIS BEE

AT

ON

Best Places to See Bees

Bees can be seen almost anywhere where there are at least a few flowers. Some of our tougher and more adaptable species, such as honey bees (see page 54) and buff-tailed bumblebees (see page 20), manage to survive even in intensively farmed land and in the centre of our large cities. The gardens and parks of suburban areas and villages can be quite rich in bee species, which will happily fly from garden to garden to visit the many different flowers, vegetable crops and weeds in the occasional unkempt or abandoned garden. Local authorities sometimes help by leaving road verges, roundabouts, portions of city parks and cemeteries unmown in spring and summer.

The countryside is a mixed bag for bees. Modern, intensive farming methods, with large fields and frequent applications of pesticides, are hostile places for most wildlife, but in contrast, some farmers take great trouble to maintain wide, flower-filled field margins and thick hedges that bloom in spring.

Small scissor bee (see page 80).

This ensures that there are sufficient bees and other insects to pollinate flowering crops such as oilseed rape, field beans and fruit orchards.

The richest sites for bees tend to be semi-natural and natural habitats such as hay meadows, chalk downland, heaths, dunes, and rides and clearings in ancient woodland. These habitats are all scarce nowadays and are often protected as nature reserves or Sites of Special Scientific Interest (SSSIs). Chalk downland is one of my favourite habitats for a bee hunt; it can be very rich in flowers and supports unusual species such as the small scissor bee (see page 80), gold-tailed melitta (see page 42) and the red-tailed mason bee (see page 78). Moorlands can also turn up some fascinating bees, such as the gloriously colourful bilberry bumblebee (see page 14).

Rare and unusual bees can also be found in unlikely places. Abandoned quarries and former factory or industrial sites can be surprisingly profitable places to search, showing that nature can be remarkably resilient and springs back given half a chance. For example, the site of a former coal-fired power station in West Thurrock, to the east of London, abandoned for 30 years, is now a wonderful place to see a range of bumblebees in abundance, as well as beautiful bees such as the bare-saddled blood bee (see page 12) and the pantaloon bee (see page 64).

Bilberry bumblebee (see page 14).

Panzer's nomad bee
Nomada panzeri

A wasp-like, slender, hairless bee, prettily patterned with chestnut and yellow stripes, Panzer's nomad is a type of cuckoo bee, parasitising the nests of tawny mining bees and some other spring-flying mining bees. Being a cuckoo, it has no need to collect pollen itself, relying instead on the host bee to perform this function for its offspring. As a result, it does not need a hairy body, as bees use hairs to brush up pollen from flowers.

Panzer's nomad bee is found throughout the UK and Ireland, and tends to be more common in the north. It is particularly associated with woodland rides, edges and coppiced areas, although in the north it is also commonly found on moorland edges.

This species flies from March to June and is most likely to be spotted hanging out near the nest burrows of its host, or grazing on nectar from cow parsley, forget-me-nots (pictured), dandelions, hawthorn or greater stitchwort.

I SPOTTED THIS BEE

AT

ON

Patchwork leafcutter bee

Megachile centuncularis

This stocky-looking solitary bee is common in gardens, and can also be found in a broad range of habitats, including scrub and woodland edge. Its favourite flowers include knapweed, thistles, burdock and brambles. The female bees carry pollen by packing it among the dense orange hairs on the underside of their abdomen.

It makes nests in holes in dead wood or crumbly masonry, and will commonly use bee hotels as long as the holes haven't already been occupied by mason bees, which fly earlier in the year. Leafcutter bees snip semi-circles of leaves to line their nest tunnel and to form a messy plug at the end, often using lilac and rose leaves which can annoy more meticulous gardeners.

The patchwork leafcutter bee is on the wing from mid-June to August, and is widespread and common in England and Wales, but scarce in Scotland and Ireland.

I SPOTTED THIS BEE

AT

ON

Red mason bee
Osmia bicornis

Probably the most familiar and abundant solitary bee species in gardens, red mason bees often nest in bee hotels, frequently in large numbers. The females have distinctive curved horns on their faces, which are used to gather, carry and mould balls of damp soil that are used to construct cells for their brood. Once a hole is filled with offspring, it is plugged rather messily with more soil. This species is most common in suburban areas, where aside from using bee hotels, it will nest opportunistically in holes and crevices in buildings, fences and other structures.

Red mason bees emerge in March, with a peak of activity in May, and a few females persisting until July. They are widespread throughout England, Wales and north to central Scotland, but absent from Ireland.

This species visits a broad range of flowers but particularly blossom of fruit trees such as apples and pears; scientists have found that red mason bees are approximately 100 times more efficient at pollinating apples than honey bees. For this reason there is some small-scale commercial breeding of red mason bees.

I SPOTTED THIS BEE

AT

ON

Red-tailed bumblebee
Bombus lapidarius

A handsome and distinctive bee, the queen and worker red-tailed bumblebees are jet black with bright red tails. The pretty males, on the wing in July and August, are additionally adorned with a bright yellow collar and fluffy yellow face, making them easy to distinguish. The Latin name derives from the old common name of this species, the stone bumblebee (*lapis* = 'stone'), after its habit of nesting under stones and in stone walls.

The red-tailed bumblebee is among our most common bees and is found in almost any flowery habitat, including gardens. It is particularly abundant on chalk downland, where it is often the most prevalent bee species. Red-tailed bumblebees occur throughout the UK but are markedly scarcer in the north.

They have a distinct preference for yellow flowers, with favourites including dandelion, bird's-foot trefoil and ragwort (pictured), but they will also happily visit knapweed and white clover.

I SPOTTED THIS BEE

AT ....

ON ....

Red-tailed mason bee
Osmia bicolor

This very pretty little bee, with a black front half and red abdomen, is unusual in that it nests exclusively in empty snail shells, using a paste made from chewed-up leaves to construct cells for its offspring. Once the nest is complete, the female bee carefully covers it with a concealing layer of dried grass stems, perhaps to avoid the attentions of cuckoo bees.

The red-tailed mason bee is among the first of the solitary bees to emerge from hibernation, with males sometimes emerging as early as late February, and defending small territories containing snail shells. Females continue on the wing until early July.

This species is very strongly associated with chalk soils, inhabiting chalk grasslands, woods and quarries. It is mainly found in the south and east of England, and can be abundant. There are a few records from Wales, but it is absent from Scotland and Ireland.

Favourite flowers include bird's-foot trefoil, kidney vetch, bluebells, ground ivy and dandelions.

I SPOTTED THIS BEE

AT

ON

Small scissor bee
Chelostoma campanularum

These tiny, slender, black bees, just 6–7 millimetres long, are easily overlooked but can be common in southern and eastern England. (They are absent from Scotland and Ireland.)

    Small scissor bees nest in pre-existing holes in wood, particularly those created by woodworm, which are frequently found in old wooden structures such as sheds and fence posts. If many holes are available, then large groupings of nests can occur. Completed nests are plugged with mud and tiny pebbles.

    As its Latin name suggests, this species will only visit flowers of the genus *Campanula* (bellflowers), such as the harebell, and in gardens they are particularly fond of the peach-leaved bellflower (*Campanula persicifolia*), shown here. Outside gardens small scissor bees are found in chalk grasslands where harebells are present. It is on the wing from June to August.

I SPOTTED THIS BEE

AT

ON

Southern cuckoo bumblebee

Bombus vestalis

This is one of the six types of bumblebee in the UK that have evolved to be social parasites of other bumblebee species. Instead of bothering to build their own nests, cuckoo bumblebee females emerge late from hibernation and seek out a nest belonging to their host species: the buff-tailed bumblebee. The cuckoo enters the host nest and attacks the resident queen, usually killing her, and then takes her place.

Given their gladiatorial lifestyle, cuckoo bumblebees tend to be large and powerful with a thick cuticle. If a cuckoo takes over, workers in the nest have little choice but to work for their new mistress, rearing her offspring and continuing to gather food for the nest. Cuckoo bumblebees do not produce workers.

The southern cuckoo can be very common in England but is scarce in the rest of the UK. The males, which are elongate and slightly spidery in appearance, are particularly fond of brambles (pictured), and in July can be seen sitting on the flowers in large numbers.

I SPOTTED THIS BEE

AT

ON

Tawny mining bee
Andrena fulva

When I was a child, we had dozens of these beautiful bees nesting in holes in the lawn of our house. The females are unmistakable, covered in dense orange-red fur, with a black head and underside. They excavate a distinctive conical spoil heap when nest-building, with the exit hole at the centre of the cone, like a miniature volcano from which the red bee erupts. The nests are often parasitised by the bee-fly *Bombylius major* (see page 102).

Tawny mining bees are on the wing from late March to June, and can be found in most habitats. They prefer to nest in sunny, short swards where rabbit grazing is intense. These bees prefer spring-flowering shrubs such as willow, blackthorn and hawthorn, and can be a useful pollinator of currant bushes and fruit trees.

The tawny mining bee is common throughout England and Wales, but rare in Scotland and Ireland.

I SPOTTED THIS BEE

AT

ON

Gardening for Bees

It is very easy to attract dozens of bee species to a garden, even a tiny outside space in an urban area. Bees are remarkably good at sniffing out flowers, wherever they might be, even on a roof terrace, balcony or window box. The important thing is to grow the right kind of flowers. Many ornamental bedding plants are hopeless; having been artificially selected for large blooms or extra petals, they have often lost their nectar or pollen, or are inaccessible to bees. Old-fashioned cottage-garden plants tend to be much better; lavender, catmint, bellflowers, geraniums (but not the related pelargoniums), lupins, borage and hollyhocks are all great choices. Herbs tend to be attractive to bees; for example chives, rosemary, thyme, mint and sage are all bee magnets. Several scientific studies have shown that pollinators generally prefer native wild flowers, so try growing some in your garden. Field scabious, viper's bugloss, marjoram, bird's-foot trefoil and foxglove are all beautiful natives that will attract bees and many other types of pollinator, such as hoverflies and butterflies. More contentiously, try to be more tolerant of native 'weeds', such as thistles, ragwort, brambles, ivy and dandelions – these are often among the favourite flowers for wild bees. A weed is just a wild flower by any other name.

Many gardens have neat, manicured lawns, which require a lot of effort to maintain and are poor for wildlife. If you have a lawn, try leaving some or all of it unmown for a while, and often bee-friendly flowers such as clovers, speedwell, buttercup and hawkbit will spontaneously appear. Ideally, from a bee's perspective, mow twice a year, once in early spring and once in late summer. If you feel that your new meadow looks untidy, mow a path through the middle and it will create the impression of order.

Aside from flowers, bees also need somewhere to nest, but the nesting requirements vary greatly between species. In gardens, bumblebees seem to have little trouble finding suitable nest sites in rodent burrows or cavities under sheds, patios and other artificial structures. Commercial bumblebee nest boxes are available, but in my experience are usually a waste of money. Bee hotels, on the other hand, often work. These provide homes for a range of cavity-nesting solitary bee species including mason bees, leafcutter bees, resin bees and yellow-faced bees. They are readily available in garden centres, or you can make your own. The basic requirement is horizontal holes, of varying diameter from

5–10 millimetres and at least 10 centimetres deep. Drilling holes in a block of wood works fine. The structure is best hung on a sunny wall or fence and, with luck, soon you will have guests.

The finishing touch to your bee-friendly patch is to be pesticide free. There is no need to use pesticides in a garden. Pesticides are poisons, designed to kill insects, plants or fungi, and most are broadly toxic to everything. You would not want to lure in bees with lovely flowers and then expose them to poison. In a healthy garden, filled with flowers and insects, there is a natural balance. Pests are rarely a problem as they are swiftly consumed by natural enemies such as ladybirds and lacewings.

Garden bumblebee (see page 40).

Tree bumblebee
Bombus hypnorum

With its distinctive banding of chestnut brown, black and white, this is one of the easiest bee species to identify. The tree bumblebee is a recent arrival in the UK from continental Europe, and was first recorded in the New Forest in Hampshire in 2001. Since then it has become common throughout England and Wales, and has spread northwards to southern Scotland, though it has not yet reached Ireland.

The tree bumblebee is unusual in that it typically nests in holes in trees well above the ground, or will enter the eaves of houses and nest under the loft insulation. They also readily adopt tit boxes, and have been known to oust nesting blue tits. In hot weather, they mass around the entrance to their nest and beat their wings to cool the inside. Tree bumblebee nests complete their life cycle in June, when males swarm around nests from which young queens are emerging.

Tree bumblebees are often common in gardens, though numbers fluctuate wildly from year to year. Keep an eye out for them on their favourite flowers, which include brambles, raspberry, cotoneaster and hardy geraniums.

I SPOTTED THIS BEE

AT

ON

Welted mason bee
Hoplitis claviventris

This is a small bee, uncommon but widespread in England and Wales, particularly the south-east. It is absent from Scotland and Northern Ireland. The female is distinctive, with a blue-black abdomen and a cream-coloured pollen brush on her underside.

Welted mason bees turn up in a range of habitats, including heaths, scrub and open woodland, flower-rich meadows, gardens and brownfield sites. They have a penchant for bird's-foot trefoil (pictured), which seems to be their main source of pollen, but they can be seen drinking nectar from buttercups, different types of scabious and the yellow flowers of plants from the daisy family (Asteraceae), such as dandelion and hawkweed.

Welted mason bees nest in the hollow stems of plants such as ragwort and brambles, the female bee first excavating the pith. Nests are sealed with a paste made from chewed-up leaves.

This species is on the wing from late May to August.

I SPOTTED THIS BEE

AT

ON

White-tailed bumblebee

Bombus lucorum

Distinguishing between white-tailed and buff-tailed bumblebees is a headache for the apprentice bee spotter. The queens are easy as the two species each live up to their name. The problem lies with the worker bees, because both species have a white tail, though buff-tailed workers usually have a hint of buff along the tail's front edge. Another way to distinguish them is that the yellow stripes of white-tails are often described as lemon yellow, and those of buff-tails as golden yellow, though the difference can be subtle. Do not fret too much however, as even professionals often lump the two together when studying them in the field.

White-tailed bumblebees are widespread in lowland UK, often occurring in gardens and feeding on a broad range of flowers including sallow catkins in spring, and later in the year on thistles, knapweed, brambles, heather, marigold (pictured), scabious and many others. This species also commonly steals nectar from deep flowers such as comfrey and aquilegia by biting a hole through the petals close to the nectary.

I SPOTTED THIS BEE

AT

ON

Willughby's leafcutter bee

Megachile willughbiella

This handsome, stocky bee may be seen on the wing from late May to August, and can be found in a range of open, sunny habitats including gardens, farmland and brownfield sites. It seems to have a particular preference for plants in the pea family (Fabaceae), such as meadow vetchling, trefoils and everlasting pea, and also for the daisy family (Asteraceae), including thistles and knapweed.

Like other leafcutter bees, nests are lined and plugged with neat semi-circles of leaves snipped from a range of shrubs and trees. This species is highly opportunistic when it comes to choosing a nest site, often occupying plant pots via the drainage holes, cracks and holes in walls, fences and window frames, and commonly using bee hotels.

Willughby's leafcutter is a common bee in southern Britain, but it is very scarce in Scotland and Ireland.

I SPOTTED THIS BEE

AT

ON

Wool carder bee

Anthidium manicatum

Its large size and yellow spots on its abdomen make this a handsome and distinctive bee. It is also unique among British bees in being highly territorial; males defend patches of flowers, attacking any insect that dares to intrude. They use mid-air headbutts or grab their foe and attempt to stab it using spines at the tip of their abdomen. Wool carder males will fearlessly attack bumblebees that may be much larger than themselves.

The females nest above ground, opportunistically using hollow plant stems, holes in dead wood and artificial structures, but rarely use bee hotels. The nest hole is lined and plugged with a woolly substance made by collecting hairs from hairy plants such as lamb's ear (*Stachys byzantina*), as shown here, or mullein (*Verbascum thapsus*). Planting lamb's ear in the garden is a reliable way of attracting this bee.

Wool carders feed mainly on woundworts and legumes such as bird's-foot trefoil, and are common in southern England and South Wales.

This species is on the wing from late May to early August.

I SPOTTED THIS BEE

AT

ON

Yellow loosestrife bee
Macropis europaea

This species has a very strong preference for flowers of yellow loosestrife (*Lysimachia vulgaris*), pictured here, and consequently is only found where this plant grows: marshes, bogs, ditches and river banks. Yellow loosestrife appears to be the only source of pollen, and also provides floral oils collected by these bees. They make their nests in damp, boggy soils but the burrows are rarely seen as they usually occur in dense vegetation. Burrows are lined with a yellow, waxy material, presumed to be made from floral oils, which is vital in preventing the nests from flooding.

This bee visits a broad range of other flowers for nectar, including thistles, knapweed, bird's-foot trefoil and brambles.

Yellow loosestrife bees are only found in southern and eastern England, and are absent from Wales, Scotland and Ireland. They fly from mid-July to the end of August, when yellow loosestrife is in flower.

I SPOTTED THIS BEE

AT

ON

Dark-edged bee-fly
Bombylius major

Many insects mimic bees, and by so doing gain some protection from predators that may be keen to avoid being stung. The dark-edged bee-fly is one such species, and is an adorable ball of flying fluff that superficially resembles a common carder bumblebee. The imitation is not entirely convincing, however, not least because the heads of bee-flies feature a long, rigid proboscis. They use this to drink nectar from flowers such as grape hyacinth, primrose, lungwort (pictured) and forget-me-not, while hovering and emitting a high-pitched buzz. Dark-edged bee-flies are on the wing from late March to early June, and are common throughout Britain, but absent from Northern Ireland. They are commonly seen in gardens, and also frequent woodland rides. Females can be seen hovering near the nests of mining bees, into which they fire their eggs with a flick of a back leg. The fly larvae are parasitoids of the bee's offspring, consuming and eventually killing them.

I SPOTTED THIS FLY

AT

ON

Bumblebee hoverfly
Volucella bombylans

Arguably our most convincing bumblebee mimic, this species commonly occurs in two colour forms: one is black with a red tail, mimicking the red-tailed bumblebee; the other has yellow and black stripes and a white tail, which mimics the white-tailed bumblebee. They often fool humans, but can readily be distinguished from real bees by their tiny, fluffy antennae.

These splendid, furry flies can be seen on the wing from May to September and are found throughout Britain and Ireland. They regularly appear in gardens and frequent woodland edges and clearings. The males often defend sunny territories, driving off rival males and welcoming females. Bumblebee hoverflies drink nectar from a range of flowers, including thistles and brambles. They lay their eggs in bumblebee nests, where the larvae scavenge on detritus, and may help to keep the nest clean. The bumblebee hoverfly not only looks like a bee but will even pretend to sting if attacked.

I SPOTTED THIS HOVERFLY

AT ....

ON ....

Recommended guides

Benton, Ted, and Nick Owens, *Solitary Bees* (William Collins, 2023)

Brock, Paul D., *A Comprehensive Guide to Insects of Britain and Ireland* (Pisces Publications, 2014)

Falk, Steven, *Field Guide to the Bees of Great Britain and Ireland* (Bloomsbury, 2015)

Goulson, Dave, *Gardening for Bumblebees* (Square Peg, 2019)

Goulson, Dave, *A Sting in the Tale* (Jonathan Cape, 2013)

Index

Note: page numbers in **bold** refer to illustrations.

alkanet, green 52
Andrena
   A. cineraria 10, **11**
   A. florea see bryony mining bee
   A. fulva 84, **85**
   A. scotica 22, **23**
Anthidium manicatum 98, **99**
Anthophora plumipes see hairy-footed flower bee
Apis mellifera see honey bee
apples 74
aquilegia 94
ashy mining bee 10, **11**
Asteraceae *see* daisy family

badgers **47**, 48
bare-saddled blood bee 12, **13**, 69
beans 40, 68
bee enemies 46–9
   see also bee-flies; cuckoo bees; cuckoo bumblebees
bee evolution 6
bee hawk-moths 8
bee hotels 9, 16, 32, 34, 52, 58, 72, 74, 88–9

bee life cycles 27–9
bee mimics 7–8, 102–5, **103**, **105**
bee-eaters 46
bee-flies 8, **8**, 47, 84, 102, **103**
beetles 7, 9
bellflower 42, 80, 87
bilberry bumblebee 14, **15**, 69, **69**
birds 46, **48**, 90
bird's-foot trefoil 14, 16, 60, 76, 78, 87, 92, 98, 100
blackthorn 10, 84
blood bees 6, 12, **13**, 69
blue mason bee 16, **17**
blue tits 90
bluebell 40, 78
Bombus
   B. distinguendus 50, **51**
   B. hortorum see garden bumblebee
   B. hypnorum 90, **91**
   B. lapidarius 76, **77**
   B. lucorum see white-tailed bumblebee
   B. monticola see bilberry bumblebee
   B. pascuorum see common carder bumblebee
   B. pratorum 38, **39**

B. terrestris see buff-tailed bumblebee
B. vestalis 82, **83**
Bombylius major see dark-edged bee-fly
borage 20, 87
bramble 12, 18, 24, 30, 34, 38, 72, 82, 88, 90, 92, 94, 100, 104
bryony mining bee 18, **19**, 42
buff-tailed bumblebee 6, **7**, 20, **21**, 67, 82, 94
buffish-mining bee 44
bugle 52
bumblebee hoverfly 8, 104, **105**
bumblebee mimics 8, 104, **105**
bumblebees 6, 8–9, 28, 30, 69, 88
  bilberry 14, **15**, 69, **69**
  buff-tailed 6, **7**, 20, **21**, 67, 82, 94
  common carder 24, **25**, 102
  cuckoo 46, 82, **83**
  early 38, **39**
  enemies 46–9
  garden 7, 40, **41**, **89**
  great yellow 50, **51**
  queens 6, 8–9, 14, 20, 28, 38, 47, 76, 82, 90, 94
  red-tailed 76, **77**
  tree 90, **91**
  white-tailed 20, 94, **95**

burdock 72
buttercup 30, 88, 92
butterflies 87

carder bees 6
  common carder bumblebee 24, **25**, 102
  wool 98, **99**
catmint 16, 20, 24, 87
cat's ear 64
chalk 42, 50, 68, 76, 78, 80
charlock 10
Chelostoma campanularum see small scissor bee
chives 87
chocolate mining bee 22, **23**
cleptoparasites 36
  see also bee-flies; cuckoo bees; cuckoo bumblebees
clover 14, 16, 24, 40, 50, 54, 60, 76, 88
Coelioxys elongata 36, **37**
Colletes hederae see ivy bee
comfrey 8, 24, 40, 52, 94
common carder bumblebee 24, **25**, 102
common furrow bee 27–8, 30, **31**
common mourning bee 32, **33**
common yellow-face bee 34, **35**
conopid flies 46, **49**
cotoneaster 38, 90
cow parsley 70

crab spiders 46
cuckoo bees 6, 9, **9**, 46
    bare-saddled blood bee 12, **13**, 69
    common mourning bee 32, **33**
    Gooden's nomad bee 44, **45**
    Panzer's nomad bee 70, **71**
    see also parasites
cuckoo bumblebees 46, 82, **83**
currants 38, 84

daisy family 34, 58, 64, 92, 96
dandelion 12, 30, 70, 76, 78, 88, 94
dark-edged bee-fly 84, 102, **103**
Dasypoda hirtipes see pantaloon bee
dead-nettle 24, 40
division of labour 29
dropwort 18
dull-vented sharp-tail bee 36, **37**

early bumblebee 38, **39**
Eucera longicornis 60, **61**
eusocial bees 27–8, 30, 62

Fabaceae *see* pea family
farming 67–8
farmland 10, 22, 24, 67, 96

fleabane 58, 64
flies 7–8, 8, **8**, 46, 47, **49**, 84, 102, **103**
floral oils 100
foraging 29
forget-me-not 70, 102
foxglove 7, 40, 87
fruit trees 32, 68, 74, 84
furrow bees 12
    common 27–8, 30, **31**
    orange-legged 62, **63**

garden bumblebee 7, 40, **41**, **89**
gardening **86**, 87–9
geranium 38, 87, 90
gold-tailed melitta 42, **43**, 68
Gooden's nomad bee 44, **45**
grape hyacinth 102
great tits 46, **48**
great yellow bumblebee 50, **51**

habitats **66**, 67–9
hairy-footed flower bee 32, 52, **53**
Halictus rubicundus 62, **63**
harebell 42, 80
hawkbit 64, 88
hawkweed 58, 94
hawthorn 10, 70, 84
heather 14, 94
herbs 87
Heriades truncorum 58, **59**

hibernation 20, 27, 28, 38, 78, 82
hogweed 10, 12, 34
hollyhock 87
honey 28, 48, 49, 54
honey bees 6, **26**, 28–9, **29**, 40, 48, 54, **55**, 67, 74
   European 49
honey buzzards 48
Hoplitis claviventris 92, **93**
hornets, Asian 49
hoverflies 87
   bumblebee 8, 104, **105**
Hylaeus communis 34, **35**

ivy 16, 52, 56, 78, 88
ivy bee 30, 56, **57**

knapweed 16, 50, 72, 76, 94, 96, 100

lacewings 89
ladybirds 89
lamb's ear 98
large-headed resin bee 58, **59**
Lasioglossum calceatum see common furrow bee
lavender 20, 24, 87
lawns 88
leafcutter bees 6, 9, 36, 88
   patchwork 72, **73**
   Willughby's 36, 96, **97**
legumes 98
   see also beans; clover; pea family

long-horned bee 60, **61**
loosestrife, yellow 100, **101**
lungwort 32, 52, 102
lupin 87

Macropis europaea 100, **101**
marigold 94
marjoram 87
mason bees 6, 9, 88
   blue 16, **17**
   red 74, **75**
   red-tailed 9, **9**, 68, 78, **79**
   welted 92, **93**
meadow vetchling 96
Megachile
   M. centuncularis 72, **73**
   M. willughbiella see Willughby's leafcutter bee
Melecta albifrons 32, **33**
Melitta haemorrhoidalis 42, **43**, 68
mining bees 6, 8–10, 44
   ashy 10, **11**
   bryony 18, **19**, 42
   buffish-mining 44
   chocolate 22, **23**
   species numbers 10
   tawny 84, **85**
mint 87
mites, parasitic 48
monkshood 8
moorlands 68–9, 70
moths 7–8, 47
mullein 98

nectar 27, 40, 46, 54, 70, 87, 92, 94, 100, 102, 104
nomad bees
    Gooden's 44, **45**
    Panzer's 70, **71**
Nomada
    N. goodeniana 44, **45**
    N. panzeri 70, **71**

oilseed rape 68
orange-legged furrow bee 62, **63**
orchid, bee 60
Osmia
    O. bicolor see red-tailed mason bee
    O. bicornis 74, **75**
    O. caerulescens 16, **17**

pantaloon bee 64, **65**, 69
Panzer's nomad bee 70, **71**
parasites 36, 46–9
    see also bee-flies; cuckoo bees; cuckoo bumblebees
patchwork leafcutter bee 72, **73**
pea family 60, 96
pears 74
pesticides 67, 89
pine resin 58
pollen 6, 27, 34, 40, 42, 46, 54, 64, 70, 72, 87, 92, 100

pollination 22, 40, 54, 60, 68, 74, 84, 87
primrose 52, 102

queens
    bumblebee 6, 8–9, 14, 20, 24, 28, 38, 47, 76, 82, 90, 94
    furrow bee 28, 30, 62
    honey bee 28–9, 54

ragwort 30, 34, 58, 64, 76, 88, 92
raspberry 18, 38, 90
red mason bee 74, **75**
red-tailed bumblebee 76, **77**
red-tailed mason bee 9, **9**, 68, 78, **79**
resin bees 88
rosemary 87
royal jelly 28

sage 87
sallows 14, 38, 94
scabious 62, 87, 92, 94
scissor bee 6
    small 68, **68**, 80, **81**
shrubs, winter-flowering 20
Sites of Special Scientific Interest (SSSIs) 68
small scissor bee 68, **68**, 80, **81**
snail shells 9, **9**, 78
social behaviour 27–9, 20, 62

solitary bees 10, 27, 46, 47, 62, 72, 74, 78, 88
Southern cuckoo bumblebee 82, **83**
speedwell 88
Sphecodes ephippius see bare-saddled blood bee
spiders 46
stings 8, 46, 48
stitchwort, greater 70
'swarming' 28

tawny mining bee 84, **85**
teasel 24
thistle 24, 34, 72, 88, 94, 96, 100, 104
thyme 87
tongue length 7, 24, 40, 50
tree bumblebee 90, **91**
trefoils 14, 16, 60, 76, 78, 87, 92, 96, 98, 100

Uists 50
umbellifers 12, 34

Varroa mite 48–9
verbena 24
vetch 50, 60, 78
viper's bugloss 87

viral diseases 49
Volucella bombylans see bumblebee hoverfly

'waggle dance' 54
wall-nesting bees 16
wasps 6
  ruby-tailed 47
wax moths 47
weeds 88
welted mason bee 92, **93**
white-tailed bumblebee 20, 94, **95**
wild carrot 34
wildflowers 87, 88
willow 10, 14, 84
Willughby's leafcutter bee 36, 96, **97**
wool carder bee 98, **99**
workers 28
  bumblebee 14, 20, 24, 28–9, 38, 76, 82, 94
  furrow bee 28, 30, 62
  honey bee 28–9, 54
woundworts 98

yellow loosestrife bee 100, **101**
yellow rattle 50
yellow-faced bees 34, **35**, 88